HEARING

by Robin Nelson

first step nonfiction

Lerner Publications Company · Minneapolis

Hearing is one of my **senses.**

I hear with my ears.

I hear loud **sounds.**
I hear a train.

I hear a jet.

I hear soft sounds.
I hear waves.

I hear a **secret.**

I hear high sounds.
I hear a bird.

I hear a **flute.**

I hear low sounds.
I hear **thunder.**

I hear a cow moo.

I hear happy sounds.
I hear laughing.

I hear music.

I hear sad sounds.
I hear crying.

I hear rain.

I hear many sounds.

What do you hear?

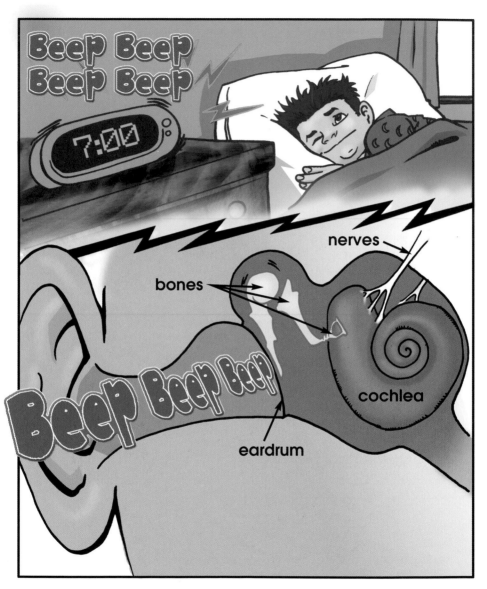

How do you hear?

When you hear something, the sound travels through the air into your ear. The sound hits your eardrum. Your eardrum vibrates. This makes three very small bones way inside your ear move. The bones send a message through a tube called your cochlea. Then the message goes through nerves up to your brain. Your brain tells you what you are hearing.

Hearing Facts

 Bats use their ears to hunt for food. They make a sound that bounces off objects and comes back to their ears as an echo. This way they can find objects in the dark. This is called echolocation.

 When you go up to the top of high mountains, your ears pop. This is caused by the change in pressure as you go higher up.

Children have more sensitive ears than adults do. Children can hear more noises than adults can.

 Dolphins have the best sense of hearing among animals. They can hear 14 times better than humans.

 Too much fluid putting pressure on your eardrum causes an earache. You might get an earache if you have an infection, allergies, or a virus.

Glossary

 flute – a musical instrument

 secret – a private thing that someone tells you that nobody else knows

 senses – the five ways our bodies get information. The five senses are hearing, seeing, smelling, tasting, and touching.

 sounds – things that you hear

 thunder – a loud noise in the air

Index

Cover image used courtesy of: © 2001 Sean Justice Productions/The Image Bank

Photos reproduced with the permission of: © RubberBall Royalty Free Digital Stock Photography, p. 2; © Trinette Reed/CORBIS, pp. 3, 22 (middle); © Betty Crowell, pp. 4, 6, 11, 22 (second from bottom); © Richard Cummins, p. 5; © Richard B. Levine, pp. 7, 22 (second from top); © Peter Johnson/CORBIS, p. 8; © 2001 D. Berry/PhotoLink, pp. 9, 22 (top); © Mark L. Stephenson/CORBIS, pp. 10, 22 (bottom); © Elaine Little/World Photo Images, pp. 12, 13, 17; © Joel Sartore/CORBIS, p. 14; © Julie Habel/CORBIS, p. 15; © Stephen G. Donaldson, p. 16.

Illustration on page 18 by Tim Seeley.

Lerner Publications Company
A division of Lerner Publishing Group
241 First Avenue North
Minneapolis, MN 55401 U.S.A.

Website address: www.lernerbooks.com

Library of Congress Cataloging-in-Publication Data

Nelson, Robin, 1971–
    Hearing / by Robin Nelson.
        p.   cm. — (First step nonfiction)
    Includes index.
    Summary: An introduction to the sense of hearing and the different
things that you can hear.
    ISBN: 0–8225–1264–5 (lib. bdg. : alk. paper)
    1. Hearing—Juvenile literature. [1. Hearing. 2. Ear. 3. Senses
and sensation.] I. Title. II. Series.
QP462.2 .N45 2002
612.8'5—dc21   2001003965

Manufactured in the United States of America
2 3 4 5 6 7 – DP – 09 08 07 06 05 04